ISBN 978-1-5283-1513-5
PIBN 10907365

THE COMPRESSIVE AND TRANSVERSE STRENGTH OF BRICK

By J. W. McBurney

ABSTRACT

This paper reports the compressive strength flat and on edge and the transverse strength of 27 makes of bricks covering a range of conditions in method of manufacture and degree of burning. The attempt is made to correlate the variation in ratios of these different measures of strength with the various structural features of the brick.

CONTENTS

I. INTRODUCTION

In a previous paper[1] the author makes the following statement: "The various measures of the strength of brick, compressive flatwise and on edge, transverse, tensile, and shearing vary in their relation one to another for different makes of brick." This statement was based upon the consideration of but five makes of bricks. The purpose of the present paper is to present data representing a range of conditions and determine if possible what are the factors influencing variation in relation of certain strength measures. A secondary purpose is that of determining the degree of justification for the practice followed in certain specifications where but one of these strength measures (usually the transverse strength) is used. Apparently it is assumed either that the various strength measures are so related that one can serve as a measure of the others or that the one selected has a sufficiently close relation to the strength of the masonry to justify the omission of the other strength tests.

II. REVIEW OF LITERATURE

Confining consideration to tests reported for the United States and Canada and taking them in chronological order the more important contributions where two or more strength tests of brick are compared are as follows:

Bleininger[2] reports 176 tests on brick collected in a district covering a radius of 150 miles from Washington, D. C. His values represent half bricks tested in compression flatwise and edgewise. He states that "the bricks tested were of the harder grade, and practically all of them were made by the stiff-mud process. * * * Both shale and clay bricks are represented." No information was provided as to whether given specimens were side cut or end cut, nor was information provided as to laminar structure. On the basis of these 176 tests, the average ratio of compressive strength flat to compressive strength on edge is given as 1.153, and "individual tests vary widely from the average ratio." From an inspection of Bleininger's graph it is evident that there is much less scattering of ratios for the clay bricks than for the shale bricks. He stated in the discussion "some of the end-cut bricks were stronger when tested on edge than when subjected to the load in the flat condition. Most of the side-cut bricks showed the opposite to be the case." Ries[3] reports results of compressive flat and on edge and transverse tests on 16 makes of bricks. These bricks are classified as dry-press, soft-mud, and stiff-mud manufacture.

[1] The Effect of Strength of Brick on Compressive Strength of Brick Masonry, Proc. Am. Soc. for Test. Materials, pt. 2, pp. 605–625; 1928.

[2] A. V. Bleininger, The Relation Between the Porosity and Crushing Strength of Clay Products, Trans. Am. Cer. Soc., 12, pp. 564–584; 1910.

[3] H. Ries, Results of Tests on Some Bricks from the Province of Western Canada, Trans. Am. Cer. Soc., 14, pp. 82–86; 1912.

Committee C-3 on Brick of the American Society for Testing Materials [4] in its report for 1915 presents results of compression tests flatwise and transverse tests of brick. Unfortunately the data as to method of manufacture are incomplete. Williams [5] reports 14 sets of tests on soft-mud and 20 sets of tests on stiff-mud bricks. These test values are for compressive strength on edge and transverse strength.

Orton [6] reports an elaborate series of comparisons between the compressive strength of half bricks tested flatwise and edgewise.

III. SOURCE OF SAMPLES

Table 1 gives values for the transverse strength, compressive flatwise strength, and compressive edgewise strength of 27 makes of bricks, arranged according to method of manufacture.

Samples Nos. 2A, 6, 13A, and 14A represent random samplings from large shipments. It will be noted that for these the number of specimens tested are 50 or more. The values for samples 7 and 17A are the averages from tests of 50 specimens selected from shipments of 300 bricks of each type. The rest of the specimens represent samples of a few bricks each supplied by the manufacturer. In all cases, however, each sample number is confined to a single grade of a given manufacturer. None of the samples represent "run of kiln."

The data of Table 1 represent tests made by the author at the National Bureau of Standards.

TABLE 1.—*Transverse and compressive strengths of bricks*

Sample No.	Number of tests	Method of manufacture	Modulus of rupture	Compressive strength, half bricks		Ratios		
				Flat	Edge	Modulus of rupture to compressive flat	Modulus of rupture to compressive edge	Compressive flat to compressive edge
			Lbs./in.²	*Lbs./in.²*	*Lbs./in.²*			
1	3	D. P.	1, 126	5, 760	5, 413	0. 195	0. 218	1. 06
2A	98	D. P.	800	3, 520	3, 630	. 227	. 220	. 97
2B	5	D. P.	141	1, 860	1, 142	. 070	. 123	1. 62
3	4	D. P.	2, 020	18, 980	11, 700	. 106	. 173	1. 62
4A	3	S. D. P.	852	7, 710	7, 220	. 111	. 118	1. 07
4B	3	S. D. P.	887	7, 300	7, 300	. 121	. 121	1. 00
4C	3	S. D. P.	293	2, 430	2, 470	. 120	. 119	. 985
5	5	S. M.	515	2, 720	3, 032	. 189	. 170	. 900
6	50	S. M.	670	3, 320	3, 360	. 201	. 200	. 990
7	50	S. M.	489	3, 110	3, 169	. 157	. 155	. 982

[4] Proc. Am. Soc. for Test. Mats., **15**, Pt. I, pp. 150–162; 1915.

[5] Ira A. Williams, Strength Tests of Oregon Building Brick, Trans. Am. Cer. Soc., **17**, pp. 660–666; 1915.

[6] Edward Orton, jr., A Comparison Between the Absorption, Crushing Strength, and Resistance to Freezing of Some Ohio Building Bricks, Trans. Am. Cer. Soc., **18**, pp. 686–760; 1916.

TABLE 1.—*Transverse and compressive strengths of bricks*—Continued.

Sample No.	Number of tests	Method of manufacture	Modulus of rupture	Compressive strength, half bricks		Ratios		
				Flat	Edge	Modulus of rupture to compressive flat	Modulus of rupture to compressive edge	Compressive flat to compressive edge
			Lbs./in.²	*Lbs./in.²*	*Lbs./in.²*			
8	3	S. M	765	3, 270	3, 630	0. 234	0. 211	0. 900
9	3	S. M	730	4, 215	3, 730	. 173	. 196	1. 13
10	1	S. M	1, 170	7, 000	7, 450	. 168	. 157	. 940
11	1	S. M	510	3, 850	4, 520	. 132	. 113	. 853
12	1	S. M	500	1, 990	2, 280	. 251	. 219	. 877
13A	100	S. M	1, 550	8, 610	11, 600	. 180	. 134	. 742
13B	6	S. M	349	1, 917	2, 240	. 182	. 160	. 856
14A	94	S. M. E. C	1, 320	3, 370	3, 780	. 392	. 349	. 891
14B	7	S. M. E. C	685	2, 710	2, 040	. 251	. 336	1. 33
15	5	S. M. E. C	1, 295	3, 040	3, 190	. 426	. 406	. 953
16A	5	S. M. E. C	947	6, 910	7, 200	. 160	. 131	. 960
16B	5	S. M. E. C	1, 250	6, 550	6, 780	. 191	. 184	. 965
16C	4	S. M. E. C	418	2, 350	2, 100	. 178	. 199	1. 12
17A	50	S. M. S. C	1, 573	10, 170	9, 104	. 154	. 173	1. 12
17B	11	S. M. S. C	776	6, 093	5, 380	. 127	. 144	1. 13
18	6	S. M. S. C	2, 130	14, 100	10, 750	. 151	. 198	1. 32
19A	5	S. M. S. C	2, 834	11, 260	10, 340	. 151	. 175	1. 09
19B	5	S. M. S. C	1, 234	13, 780	10, 440	. 090	. 117	1. 33
20	2	S. M. S. C	1, 477	8, 085	7, 415	. 183	. 200	1. 09
21	25	S. M. S. C	2, 890	22, 600	18, 950	. 128	. 153	1. 19
22A	5	S. M. S. C	1, 864	5, 950	5, 560	. 314	. 335	1. 07
22B	5	S. M. S. C	1, 170	4, 040	4, 070	. 290	. 288	. 944
23A	2	S. M. S. C	603	8, 100	3, 585	. 074	. 168	2. 26
23B	3	S. M. S. C	721	5, 990	3, 816	. 120	. 189	1. 57
23C	3	S. M. S. C	307	3, 546	2, 030	. 087	. 151	1. 75
24A	3	S. M. S. C	3, 230	21, 200	14, 200	. 152	. 227	1. 49
24B	4	S. M. S. C	987	6, 250	5, 890	. 158	. 168	1. 06
25	1	S. M. S. C	1, 320	7, 940	9, 500	. 166	. 139	. 835
26	1	S. M. S. C	1, 020	6, 570	5, 750	. 155	. 177	1. 14
27	1	S. M. S. C	790	3, 270	3, 470	. 242	. 228	. 942

D. P.=Dry press.
S. D. P.=Semidry press.
S. M.=Soft mud.
S. M. E. C.=Stiff mud, end cut.
S. M. S. C.=Stiff mud, side cut.

IV. METHODS OF TESTING

The tentative methods of testing brick (C67–27T) of the American Society for Testing Materials were followed. The half bricks from the transverse test were used in the compressive tests, one of the two halves from each brick being tested flatwise and the other on edge. Where the transverse tests gave a break which was oblique or otherwise irregular, the halves were trimmed by cutting or grinding.

V. DESCRIPTION OF TEST SPECIMENS

1. DRY PRESS BRICKS

1. Well made and free from lamination. Sample supplied by manufacturer and represented "well-burned" color range, dark to light.

2A. Random samples from shipment of 52,000 "commons." Granular texture, no laminations. Color reddish brown, surface clay.

2B. Samples supplied by the manufacturer of 2A, but quite underburned salmons. Weak and friable. Color pink to orange.

3. This sample represents an experimental dry press brick formed and burned at the plant of a brick-machinery manufacturer. Nos. 19A and 19B represent the same shale and, in the opinion of the plant superintendent, 19A represents the same degree of vitrification as No. 3.

2. SEMIDRY PRESS BRICKS

4A. "Arch," or "clinker." Very considerably cracked and crazed. The cracks were apparently kiln cracks, since samples 4B and 4C were comparatively free from them. Some warpage.

4B. "Body" brick corresponding to 4A. Much less evidence of kiln cracking. Good red in color.

4C. The salmon of this manufacture. Orange in color.

3. SOFT-MUD BRICK

5. Manufacturer's sample representing a considerable range of color. Some kiln cracking and warping.

6. Random sample representing 18,000 "select common"; occasional nodules. The harder burned bricks (as judged by color) frequently showed kiln cracks normal to the long axis of the brick.

7. Sample represents random sampling of shipment of 300 "select common." Brick characterized by a sandy core and numerous lime nodules in size up to three-quarters inch in diameter. It may be of interest to note that salmons corresponding to No. 7 were provided, but expansion of the nodules disrupted the specimens during drying. About one-third of specimens tested showed cracks.

8. "Typical common brick of district" in judgment of building inspector. Resembles No. 6 in appearance. Some irregular laminar structure.

9. Also considered "typical common brick." Very fine grained. Some small nodules. No noticeable laminar structure.

10. One specimen. Nodules and cavities evident on examining cross section.

11. One specimen. Gave a very oblique break when tested transversely.

12. One specimen. Nothing noted in way of lamination, nodules, or cavities.

13A. Sample from shipment of 52,000 brick. It was impossible to distinguish any granular or laminated structure. Appearance as of a fused body. Core colors different from outer portion of brick were quite common. Colors dark.

13B. Salmons of 13A light orange. Weak. Original clay grains easily distinguishable.

4. STIFF MUD, END CUT

14A. Double column. Due to lamination gave the appearance of being made of bundles of fibers running parallel to long axis of brick. Frequently on compression a circular core extending the length of the brick would result from spalling of the sides. Black cores were common. The brick itself, aside from its structure, gave some evidence of being rather hard. Large lime nodules were common.

14B. The "unmarketable salmon" of 14A. There was very little to distinguish it in appearance from 14A. Noticeably weaker.

15. A manufacturer's sample. Sample 15 is indistinguishable in appearance and properties from 14A. It, however, comes from a different district.

16A. "Rough hard," which is the local name for "the arch" or "clinker" bricks of other districts. This brick showed much cracking, crazing, and warping, probably due to overburning. Little evidence of lamination. Such lamination as existed gave planes parallel to the broad face of the brick.

16B. The "straight hard" or "body" brick corresponding to 16A. Good red and free from cracking and crazing characterizing 16A. Slight lamination as in 16A.

16C. The salmon of the two preceding samples. Color was characteristic of that which gave the name salmon. No cracking or crazing and otherwise resembled 16B.

5. STIFF MUD, SIDE CUT

17A. Random sample of 50 from manufacturer's sample of 300. Well-made shale. Granular structure, no evidence of lamination. The entire sample was remarkably uniform in all its properties.

17B. The salmon of 17A. Resembles 17A except for color and strength.

18. Well-made characteristic shale. No evidence of lamination.

19A. Manufacturer's sample. Same shale and, in opinion of manufacturer, same burning history as sample No. 3. No evidence of lamination.

19B. Another sample from maker of 19A. From evidences of color and comparative water absorption 19B was not as hard burned as 19A.

20. Manufacturer's sample "common." No lamination evident. Shale brick, but from color and water absorption apparently not very hard burned

21. Manufacturer's sample of "hard shale commons." No laminations. Degree of burning such that granular structure was practically suppressed.

22A and 22B. Together represent a vertical section of a down-draft kiln; 22A is the top half, 22B the lower half. From color and water absorption the grading is obvious. The structure is granular, with an appearance of what might be called internal crazing; small short cracks at irregular intervals and direction. Shale.

23A. Manufacturer's sample of "clinker." Black glazed exterior, red interior. This and samples 23B and 23C were outstanding examples of die lamination. Apparently an imperfect mixture of two clays existed, and the laminar structure was very noticeable. Aside from lamination, sample 23A showed the cracking, crazing, and warping usually associated with clinker bricks.

23B. The body brick corresponding to 23A. Highly die laminated, not so irregular as 23A.

23C. The salmon corresponding to the two preceding samples. Also laminated but little cracked.

24A. Manufacturer's sample offered as "face." These represent the top portion of the kiln. Hard burned, unlaminated shale.

24B. Lower portion of kiln. Marked "fillers" by manufacturer. Salmon color; otherwise description of 24A applies.

25. One specimen side-cut clay. Some nodules, no lamination evident.

26. One specimen. Some irregular lamination (auger), clay.

27. One specimen. Clay, sandy texture, no lamination.

VI. DISCUSSION OF RESULTS

1. RELATION OF COMPRESSIVE STRENGTH FLATWISE TO COMPRESSIVE STRENGTH EDGEWISE

Assuming that brick behaves like the Swiss sandstone tested by Bauschinger,[7] the ratio of compressive strength flatwise to compressive strength edgewise would be 1.26. Actually, the data here presented show a wide variation. The two extremes for this ratio are 0.74 for sample No. 13A and 2.26 for sample No. 23A. It is obvious that the structure of the brick is responsible for this variation, hence it is in order to examine these data for the purpose of seeing if there is any correlation between the structure of the brick and the ratio of compressive strength flatwise to compressive strength edgewise.

[7] Johnson's Materials of Construction, 6th ed., pp. 113–114.

Referring to the descriptions of the samples, it will be observed that the following data have been recorded:

(*a*) Method of manufacture.

(*b*) Kind of raw materials, clay or shale.

(*c*) Presence or absence and character of laminations and cracks.

(*d*) Presence or absence and character of nodules or other inclusions.

(*e*) Character of texture; that is, granular, glassy, etc.

(*f*) In some cases degree of burning is noted.

In the opinion of the writer too much weight should not be given to differences within 10 per cent where the samples are small, due to the large variation in results due to sampling.

(*a*) Effect of Method of Forming.—Considering the effect of method of forming on the ratio compressive strength flatwise to edgewise, it would appear on the basis of the very limited series of dry-press specimens that the range for this ratio is from unity up to 1.6. Increase in the ratio appears to be produced by pronounced underburning (note sample 2B in comparison with 2A). No obvious explanation is noted for the high ratio of No. 3. It has been observed that excessive pressure in forming a dry-press brick will produce laminations or planes of weakness normal to the direction of the pressure. The results of Ries [8] give a range of from 1.06 to 1.72 for five samples of dry-press bricks.

The semidry press samples are close to unity for their ratio. The fact that the arch brick (4A) is somewhat stronger flat than on edge would seem explainable by its cracked and crazed structure.

The soft-mud samples, with one exception (No. 9), were all stronger on edge than flat.

The eight samples of soft-mud bricks reported on by Ries [8] were stronger on edge than flatwise. The five samples reported by Orton [9] gave ratios varying from 0.93 to 1.18 for the hard and medium burns. It is evident from the data here presented that soft-mud bricks tend toward a lower ratio than characterizes the other methods of manufacture. Why this should be is not apparent. Soft-mud bricks are formed by light pressure in a mold at semiliquid consistency. No laminar structure or planes of weakness should be formed. They should show the effect of change in ratio of cross section to a height to give greater strength flatwise than edgewise. The only explanation that comes to mind is that these bricks are burned on edge and the pressure of the bricks may produce a consolidation in the line of the applied forces giving greater strength on edge than on flat. This possibility will be mentioned again in connection with the discussion of effect of degree of burning on the compressive-strength

[8] See footnote 3, p. 822. [9] See footnote 7, p. 827.

ratio. Like the soft-mud bricks, the well-burned stiff-mud end-cut bricks were generally stronger on edge than on flat. The salmon stiff-mud end-cut show the reverse of this.

The characteristic of side-cut bricks is that of being stronger flat than on edge. In general, the data of Orton and Ries confirm this.

(*b*) Effect of Kind of Raw Material.—It is not considered that these data are sufficient to warrant any conclusions as to differences between clays and shales which may affect the ratio of compressive strengths flat to compressive strength on edge. It should be remembered that the type of raw material is usually the consideration that determines the method of forming.

(*c*) Presence and Character of Laminations and Cracks.—Considering the same clay molded by different processes, the presence or absence of laminations in its various degrees is largely a result of

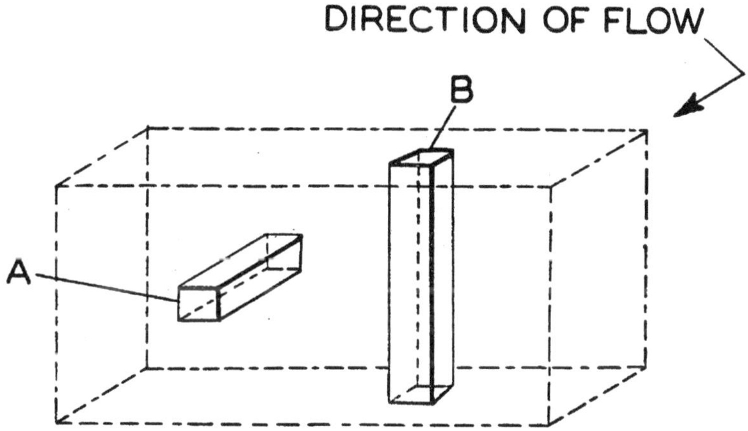

Fig. 1

the methods of manufacture. Soft-mud bricks should be practically free from lamination. Stiff-mud bricks may be substantially free from laminations or may have a highly laminar structure. Dry-press bricks may develop planes of weakness normal to the direction of the applied pressure, when improperly pressed.

Considering a laminated side-cut brick, the flow of material in the die is normal to the large face. Hence, the planes of lamination would, in general, be normal to the large face. It would appear reasonable to expect that a small prism, such as *A* in Figure 1, would have a greater strength than *B*, due to the compacting of the grains in the case of prism *A* and their separation due to differential flow in the case of prism *B*. It is conceived that burning on edge may have an effect toward overcoming this tendency, which may explain the occasional reversal of the ratio. The presence of irregular cracking and crazing, such as frequently characterize clinker bricks, would seem to offer an explanation of high ratios of flat to edge compressive strength. Surface cracks would be expected to have a greater effect on the compressive strength when the bricks were tested on edge, because the portions cracked would occupy a greater percentage of the sectional area than when the bricks were tested flatwise.

(*d*) EFFECT OF NODULES.—Presence or absence of nodules does not appear to have any definite effect on the relation between the two compressive strengths. However, it is possible that if a nodule, inclusion, or core were of a size such that it represented a considerable portion of the cross section of the brick and were of a different strength, then an effect would be produced on the relative compressive strength flatwise to edgewise by the difference in percentage of the two materials effective for the two cross sections. Figure 2 shows schematically a cross section normal to the long axis of sample No. 13A. The core in this case is due to lack of complete oxidation. Assuming that the outer portion *A* is stronger than the inner portion *B*, it is evident from the diagram that a greater percentage of the cross section along the line *x*---*x* is of material *A* than is the case for the cross section *y*---*y*. If this difference in strength exists, this would appear to explain the low ratio (0.742) for sample 13A.

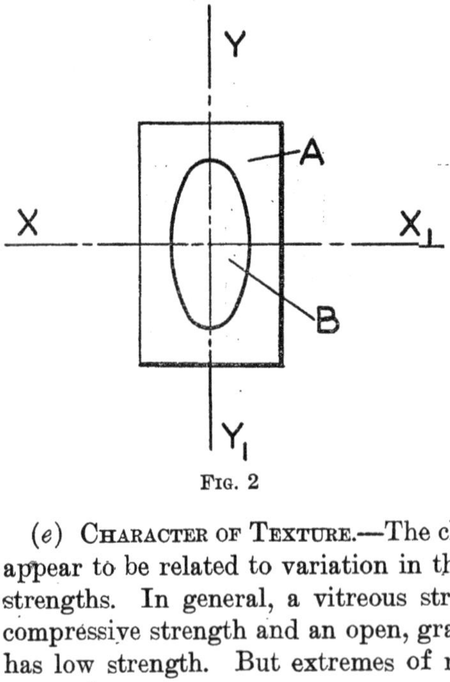

FIG. 2

(*e*) CHARACTER OF TEXTURE.—The character of the texture does not appear to be related to variation in the ratio of the two compressive strengths. In general, a vitreous structure is associated with high compressive strength and an open, granular, poorly bonded structure has low strength. But extremes of ratios are associated with both types of structures.

(*f*) DEGREE OF BURNING.—A casual inspection of the data where the product of the kiln is classified into arch, body, and salmon, or where the sample represents a vertical section of the kiln and classification was made by position, seemed to indicate a very marked effect on ratios of compressive strength by degree of burning. However, in some cases (2A and 2B, 13A and 13B, 14A and 14B, 16A, 16B and 16C) the effect was for the underburned brick to have higher ratios of compressive strength flat to edge. In other cases (22A and 22B, 24A and 24B) the direction of change was reversed. In yet other cases there was no significant difference in the ratios (4A, 4B, and 4C, 17A and 17B) or the order was mixed (23A, 23B, and 23C).

However, the suggestion that the burning of bricks on edge might add to their apparent compressive strength on edge by the effect of the weight of the superimposed bricks during burning does much to explain these data. Samples 2, 13, 14, and 16 represent bricks from scove kilns. The salmons from scove kilns come from the top and outer portions of the kilns. They represent specimens that have been subjected to relatively little pressure in burning. Samples 22 and 24 are from down-draft kilns. Here the salmons are from the bottom of the setting. Samples 4 and 16 are from small scove kilns, and the clinker specimens are both quite cracked and crazed. No. 17 was burned in a down-draft kiln, but the location of the specimens is unknown. The type of kiln used for No. 23 is also unknown. In any case the higher ratio of the clinker brick is explainable by its cracked and crazed condition.

(*g*) The Character of the Fracture in Compressive Tests Edgewise.—Before leaving the subject of tests in compression it is desired to introduce some data dealing with the apparent relation between compressive strength and type of fracture.

Table 2 gives an analysis of the data secured on a test of 50 specimens of brick 2A, the samples being whole brick tested edgewise. The type of fracture, as indicated on the end of the test specimen, was recorded at the time of test.

Table 2 gives clearly the fact that a double shear (cone or wedge) is associated with a higher compressive strength than where the failure takes place in a single diagonal shear. This fact alone could be explained by assuming that the 10 tests which gave the higher strength and double shear were concentrically loaded, while the 13 tests which failed in single diagonal shear with low strength were eccentrically loaded. However, the fact that the difference in strength is actual and not due to difference in loading is vouched for by the inverse relation of the water absorption. The possibility exists that eccentric loading was present for both types of fractures, but the stronger bricks were better able to resist the effect of eccentricity.

Table 2.—*Brick 2A*

[Total number of specimens, 50]

Number of tests	Compressive strength, whole brick on edge	Water absorption, five-hour boil	Type of break (see fig. 3)
	Lbs./in.²	*Per cent*	
50	3,300	22.26	All types.
10	4,100	21.29	A.
13	2,470	23.95	B.
27	3,400	21.80	C, D, E, F.

2. RELATION OF TRANSVERSE TO COMPRESSIVE STRENGTH

The ratios of modulus of rupture to compressive strength flat given in Table 1 range from 0.426 to 0.070. The transverse strength is notably quite sensitive to imperfections in the brick. Usually it will be found that where a series of bricks are tested the percentage deviation from the mean of the modulus of rupture will exceed the percentage deviation from the mean of the compressive strength. In other words, nodules, slight laminations, or other planes of weakness will very noticeably reduce the transverse strength.

(*a*) EFFECT OF METHOD OF MANUFACTURE.—As would be expected, soft-mud bricks were the most constant in this ratio. Nodules are the principal interfering factor likely to be present. The 10 ratios

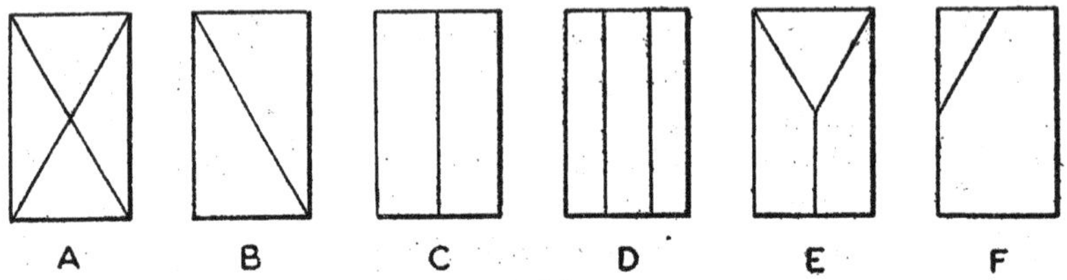

FIG. 3

given in Table 1 range between 0.132 and 0.251. Fourteen values given by Williams[10] range from 0.171 to 0.256.

The ratios for dry-press bricks range from medium values, 0.227, down to 0.070.

The ratios for stiff-mud end-cut bricks range from medium values to 0.426.

For stiff-mud side-cut bricks the ratios differ rather widely, going from 0.074 to 0.314.

(*b*) EFFECT OF TYPE OF RAW MATERIAL.—If there is any consistent effect on the relations between transverse and compressive strength due to the use of shale or clay, it is obscured by the variation in other factors.

(*c*) EFFECT OF LAMINATIONS AND CRACKS.—Not only do laminations exercise the obvious effect of lowering the ratio modulus of rupture to compressive strength, but in certain cases they are believed to be responsible for raising the ratio. Samples 14A and 15 are considered examples of this effect. They are both double-column, end-cut bricks and were auger laminated to a very considerable degree. The net result was to produce a tubular structure. From analogy to the strength of a metal tube acting as a beam and the strength of a metal tube in compression normal to its axis as compared to the corresponding strengths of a solid shaft of the same diameter, this explanation of the high ratio of transverse to com-

---

[10] See footnote 5, p. 828.

pressive strength for these bricks seems warranted. Samples 23A, 23B, and 23C are illustrations of laminar structure acting to lower this ratio.

Kiln cracking normal to the long axis of the brick is, of course, a sure way to lower the transverse strength.

(*d*) Effect of Nodules.—From observations of the tests of soft-mud bricks it is believed that a heterogeneous section, the visible evidences of which are nodules, has much to do with moderate variation in the ratio nodulus of rupture to compressive strengths. The effect of texture is uncertain.

(*e*) Effect of Variation in Burning.—No definite effect on modulus of rupture—compressive-strength ratio seems to be produced by variation in burning. The salmon bricks are usually not far from the well-burned bricks in ratio. Samples 2A and 2B provide a notable exception. The clinker bricks are usually low in ratio, probably explainable by their cracking and crazing.

VII. SPECIFICATION REQUIREMENTS

In the words of Orton, "We can not safely translate data made by crushing on the flat into terms of crushing on the edge, except in large masses or averages, and here with many reservations."

These data, in the writer's opinion, give no justification for the belief hold in some quarters that the B grade of the Tentative Specification for Building Brick (C62–27T) of the American Society for Testing Materials is necessarily the equivalent of the medium grade of the former Standard Specification for Building Brick (C21–20) of the same society. Grade B (C62–27T) gives compressive strength flatwise as 2,500 to 4,500 lbs./in.^{2} Grade medium (C21–20) gives compressive strength edgewise as 2,000 to 3,500 lbs./in.^{2} The equivalence of these two grades is certainly not true for soft-mud brick and only occasionally fit certain kinds of stiff-mud bricks.

The data on transverse strength compared with compressive strength likewise gives no warrant for assuming any general relation between the measures of strength such as would be expected for isotropic and homogenous material. The fact that a given ratio may be the average for a very large number of tests does not increase the probability that the average ratio will represent the actual ratio for a particular kind of brick. As an extreme illustration, consider a field occupied by an equal number of cattle and horses. The statement that the animals in that field have on an average one horn apiece illustrates the fallacy of attempting to average separate categories. The practice of plotting one measure of strength against another measure of strength where a wide scattering exists, as in the present data, and then drawing a line by least squares to represent the "most probable" relation would be valid if the scattering of the points represented "error." To illustrate: Sample 14A is typical of that clay and method of manufacture. Repeated tests confirm its high ratio (0.392) quite closely. The fact that brick 14A has a high modulus of rupture along with a comparatively low compressive strength is not chance. It is the nature of that brick. Hence, taking a ratio based on averages of all bricks does not represent the truth concerning this particular kind of brick.

The application of this to specification writing is that brick is purchased as the product of a given manufacturer. If a given property (say a certain flat compressive strength) is desired, that property should be asked for, not another property assumed to be related to the wanted property by a presumably fixed ratio.

VIII. SUMMARY

From the data and discussion here given the following conclusions are believed warranted in so far as they are limited to bricks made of shale or clay.

1. The ratio compressive strength of brick flat to compressive strength of brick on edge ranges from 0.74 to 2.3.

2. The tendency of soft-mud brick is to give higher unit strengths tested on edge than when tested flat.

3. The "compacting effect" on the structure of the edge-set brick by the superimposed weight of the other bricks in the kiln is offered as a tentative explanation of the tendency toward higher strength on edge.

4. This tendency toward higher strength on edge is overcome by laminar and cracking structure in the case of certain bricks.

5. The ratio of modulus of rupture to flat compressive strength ranged from 0.426 to 0.070.

6. Soft-mud bricks tend to display less deviation in the ratio modulus of rupture to flat compressive strength than any of the other methods of manufacture, but even with these the ratio ranged between 0.13 and 0.26.

7. Auger lamination in end-cut brick appears to be associated with high ratios for modulus of rupture to flat compressive strength.

8. Die lamination in side-cut brick appears to be associated with low ratios for modulus of rupture to flat compressive strength.

9. In view of the variation in the ratio compressive strength flat to compressive strength on edge, there exists no general rule for converting values from one kind of test to the other kind.

10. In view of the variation in the ratio modulus of rupture to compressive strength flat, the possibility of inferring a compressive strength from a transverse test or vice versa is open to very large errors for any given make of brick.

IX. ACKNOWLEDGMENTS

The author acknowledges his indebtedness to the various members of the staff of the National Bureau of Standards who gave assistance and counsel in the preparation of this paper. Particular thanks are due to R. T. Stull for his suggestions drawn from his plant experience.

Acknowledgment is made of the assistance rendered by the various manufacturers who generously provided samples and information.

WASHINGTON, November 23, 1928.

Bureau of Standards Journal of Research

PARTIAL CONTENTS OF VOLUME 1—Continued

October, 1928 (Vol. 1, No. 4)—Continued

November, 1928 (Vol. 1, No. 5)

December, 1928 (Vol. 1, No. 6)

Research papers are available as separates for purchase from the Superintendent of Documents
United States Government Printing Office, Washington, D. C.

[Continued on page 4 of cover]

Bureau of Standards Journal of Research

CONTENTS OF VOLUME 2

January, 1929 (Vol. 2, No. 1)

February, 1929 (Vol. 2, No. 2)

March, 1929 (Vol. 2, No. 3)

Research papers are available as separates for purchase from the Superintendent of Documents
United States Government Printing Office, Washington, D. C.

CPSIA information can be obtained
at www.ICGtesting.com
Printed in the USA
BVHW090434201118
533516BV00014B/959/P